Neptune

Carmen Bredeson

Franklin Watts
A Division of Scholastic Inc.
New York • Toronto • London • Auckland • Sydney
Mexico City • New Delhi • Hong Kong
Danbury, Connecticut

For Larry Bredeson, my star-gazing companion

Note to readers: Definitions for words in **bold** can be found in the Glossary at the back of this book.

Photographs © 2002: Corbis Images/Roger Ressmeyer: 17, 22, 24; NASA: 49 (J. Elliot/MIT), cover, 27, 28, 34, 37, 43, 51, 52 (JPL), 3 left, 4, 8, 14, 26, 30, 35, 38, 41, 48; Photo Researchers, NY: 16 (Luke Dodd/SPL), 3 right, 54 (Mark Marten/US Geological Survey), 32 (Seth Shostak/SPL), 10, 12 (SPL), 44 (Detlev Van Ravenswaay/SPL); Photri Inc.: 15 (Brent Winebrenner), 7, 46.

Solar system diagram created by Greg Harris

The photograph on the cover shows a 1989 *Voyager 2* image of Neptune.

Library of Congress Cataloging-in-Publication Data

Bredeson, Carmen.
    Neptune / Carmen Bredeson.
       p. cm. — (Watts library)
    Includes bibliographical references and index.
    Summary: Describes the discovery, exploration, atmosphere, geography, and moons of Neptune.
    ISBN 0-531-12037-6 (lib. bdg.)   0-531-16615-5 (pbk.)
    1. Neptune (Planet)—Juvenile literature. [1. Neptune (Planet)] I. Title. II. Series.

QB691 .B74 2002
523.48'1—dc21

2001007414

©2002 Franklin Watts, a Division of Scholastic Inc.
All rights reserved. Published simultaneously in Canada.
Printed in the United States of America.
1 2 3 4 5 6 7 8 9 10 R 11 10 09 08 07 06 05 04 03 02

Contents

Chapter One
Journey to Neptune 5

Chapter Two
A New World 11

Chapter Three
***Voyager 2* to the Rescue** 23

Chapter Four
Up and Over 33

Chapter Five
The Voyage Continues 45

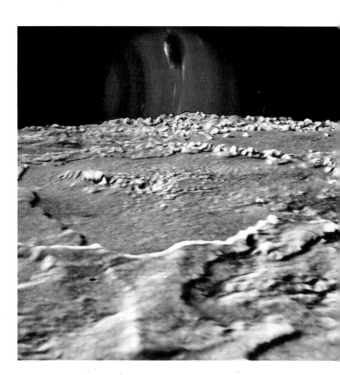

55 **Glossary**

57 **To Find Out More**

60 **A Note on Sources**

61 **Index**

Voyager 2 *was launched from the Kennedy Space Center in Florida on August 20, 1977.*

Chapter One

Journey to Neptune

For twelve years, beginning in 1977, *Voyager 2* traveled through the blackness of space. Images sent back from the little space probe during its journey gave people on Earth their first close-up look at Jupiter, Saturn, and Uranus. The last stop on *Voyager 2's* tour was Neptune, the fourth-largest planet in the solar system. By the time the space probe arrived at Neptune, it would have traveled nearly 4 billion miles (6.4 billion kilometers) through space.

Would *Voyager 2's* cameras still be working after the probe's long journey? Would the engineers on Earth be able to aim them in the right direction? During the early days of 1989, scientists at the Jet Propulsion Laboratory (JPL), part of the National Aeronautics and Space Administration (NASA), started getting nervous. It was about time for data to begin arriving from *Voyager 2*. This was the last stop on a very successful mission, and all the scientists had their fingers crossed that things would go well.

Voyagers in Space

There were two Voyager spacecraft sent to the outer planets. Scientists wanted to be sure to have a backup in case something happened to one of the probes. *Voyager 2* was launched on August 20, 1977, followed by the launch of *Voyager 1* on September 5, 1977. Each probe weighs about a ton, is the size of a small car, and looks like a big satellite dish. The Voyagers carry an array of cameras, computers, and scientific instruments to record their encounters in space.

When launched in 1977, both probes were headed to a rendezvous with Jupiter and Saturn. In 1979, they arrived at Jupiter and sent back amazing pictures of the planet and its moons. They then flew by Saturn and returned images of that planet's extraordinary rings.

After the Saturn flyby, *Voyager 1* headed on a path that will eventually take it out of the solar system. This spacecraft will drift in space forever unless it crashes into another solid

Gas Giants

Jupiter, Saturn, Uranus, and Neptune are known as gas giants. The centers of these huge planets are made up of molten rock that is surrounded by a gas envelope. If you were able to "land" on one of the gas giants, you would have no solid surface to stand on.

A prototype Voyager spacecraft undergoes vibration tests at NASA's Jet Propulsion Laboratory in Pasadena, California.

object. NASA decided to continue the journey of *Voyager 2* by sending it on to Uranus and Neptune. These two outer planets are relatively recent discoveries in the history of astronomy. Scientists did not know that Uranus existed until 1781, and it took another fifty-five years for Neptune to be located in the vastness of space.

Finding Uranus

William Herschel was a British amateur astronomer who built telescopes in his kitchen. Herschel and his sister, Caroline, decided to make a detailed map of the stars. Night after night, Willliam searched the sky while Caroline took notes and drew diagrams of what her brother saw. On March 13, 1781, William saw something that he thought was a **comet** at first,

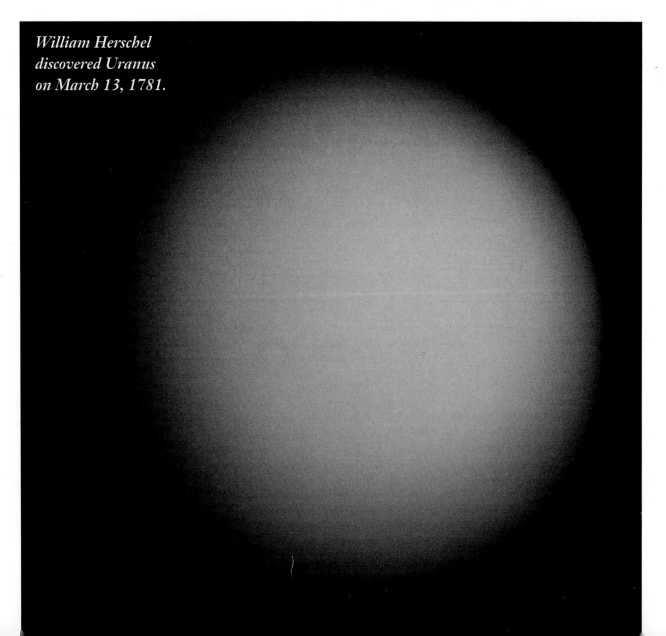

William Herschel discovered Uranus on March 13, 1781.

> **Sky Search**
>
> From our point of view on Earth, stars appear to stay in the same position in the sky in relation to each other. Planets, on the other hand, **orbit** the Sun, so they continually move across the background of the star patterns.
>
> When astronomers look for **asteroids**, comets, planets, and other celestial bodies, they compare star charts taken over several days or weeks. They scan the charts for anything that changes position in the star patterns.

but it was moving too slowly to be a comet. After more study, he decided it was a planet—the first one discovered in modern times. The new object was eventually named Uranus, after the Greek god of the sky.

After Uranus was discovered in 1781, astronomers noticed that it wobbled a little in its journey around the Sun. Some thought that another planet with a strong gravitational pull might be causing the wobble. But where was the elusive planet?

English astronomer John Couch Adams calculated the position of an undiscovered eighth planet beyond Uranus.

Chapter Two

A New World

John Couch Adams, a twenty-three-year-old mathematics student at Cambridge University in England, was interested in finding the planet that seemed to affect the orbit of Uranus. He wrote in his journal on July 3, 1841, "Formed a design in the beginning of this week of investigating, as soon as possible after taking my degree, the irregularities in the motion of Uranus . . . in order to find out whether they may be attributed to the action of an undiscovered planet beyond it."

Adams got his degree in 1843 and soon began calculating the location of the

French astronomer Urbain Jean Joseph Le Verrier calculated Neptune's position at the same time as Adams.

mystery planet. By September 1845, he thought he had its position pinned down. Adams did not publish his results, but he gave the information to the director of the Cambridge Observatory, James Challis. The calculations were then sent to astronomer Sir George Airy at England's Royal Observatory.

Airy did not investigate Adams's claims. During September and October 1845, Adams tried to visit Airy three times. Adams had not made an appointment in advance, and Airy was away for two of the visits. Airy apparently did not get the message that Adams was there on the third visit. Adams, who was a very shy man, did not try to see Airy again. When Airy finally wrote to Adams asking for additional information, Adams did not answer the letter for a year. One of Adams's associates at Cambridge said that Adams "acted like a bashful boy rather than like a man who had made a great discovery."

Meanwhile, at about the same time, French astronomer Urbain Jean Joseph Le Verrier was also searching for a planet that might be affecting the orbit of Uranus. By June 1846, he had pinpointed a proposed location for the eighth planet. Le Verrier published his findings and sent his calculations to the French Academy. He also wrote to Johann Gottfried Galle, a German astronomer at the Royal Observatory in Berlin.

After seeing Le Verrier's calculations, Galle immediately became interested in the project. Hours after receiving the letter on September 23, 1846, Galle sat down at one of the observatory telescopes. His assistant, Heinrich d'Arrest, suggested that Le Verrier use a recent star chart to help in the search. As Le Verrier called out the locations of the stars he observed, d'Arrest found them on the chart.

This Is a Big Fellow!

At about midnight, Galle found a star that did not appear on the chart. D'Arrest cried out, "That star is not on the map!" What Galle saw was not a star at all, but a new planet. He switched his telescope to high power to get a better look and exclaimed, "My God in heaven, this is a big fellow." D'Arrest took a look and said, "There it is! There is the planet, with a disk as round, bright, and beautiful as that of Jupiter!"

Galle wrote to Le Verrier on September 25, 1846, stating, "Monsieur, the planet of which you indicated the position really exists." Le Verrier replied, "I thank you for the alacrity with which you applied my instructions. We are thereby, thanks to you, definitely in possession of a new world."

On October 1, 1846, the *London Times* announced, "Le Verrier's Planet Found." Both Adams and Le Verrier were

This is about how Neptune (right of center) and Triton (upper left of center) would have appeared to Gottfried Galle.

right on target. Using only mathematics and hand calculations, they had predicted locations extremely close to each other. Galle suggested naming the new planet Janus after the Roman god of doors and boundaries. Le Verrier did not like this name because it might imply that the planet lay at the boundary, or end, of the solar system. They finally settled on Neptune, the Roman god of the sea.

Eyes on the Sky

Ancient astronomers followed the paths of Mercury, Venus, Mars, Jupiter, and Saturn before telescopes were invented. These planets are large and bright enough to be seen with the naked eye. Neptune is too far away and too dim to be seen without the aid of a telescope. After Neptune was discovered, telescopes all over the world were aimed at the new planet.

On October 10, 1846, amateur astronomer William Lassell was looking through his homemade telescope in Liverpool, England. As he studied Neptune, he noticed a bright dot "whose situation . . . occasioned my strong suspicion that it may be a satellite." In July 1847, after more observations,

Neptune was named after the Roman god of the sea, shown here as a sculpture at the Palace of Versailles in France.

Billions of Stars

The Milky Way Galaxy is made up of more than 400 billion stars. The Sun is just one of those stars. The universe is made up of billions of galaxies that are similar to our Milky Way.

Lassell's suspicions were confirmed. Neptune's new moon was named Triton after the son of Poseidon, Greek god of the sea.

Astronomers could not see any details on either Neptune or Triton because of their great distance from Earth and the Sun. They continued to study both, however, to try and figure out how big they were. They also wanted to learn how long Neptune takes to orbit the Sun. Finally, were there any other moons circling Neptune?

Patterns of Light

As scientific equipment continued to improve, astronomers began learning more and more about the newest planet. They analyzed light from Neptune with a **spectrograph**. This type of scientific instrument, which can be attached to a telescope, breaks up light and separates it into different colors. Each chemical emits, or gives off, its own color. By analyzing the bands of color, astronomers can draw a picture of the chemical makeup of an object. Early spectrographic studies of Neptune showed that its **atmosphere** is made up of **hydrogen**, **helium**, and **methane**.

It is the methane in Neptune's atmosphere that gives the planet its blue

A scientist hooks up a spectrograph at Las Campanas Observatory in Chile.

color. Methane absorbs the wavelengths of sunlight near the red end of the color **spectrum**. The remaining colors from Neptune's atmosphere are those near the blue end of the spectrum. The blue colors are not absorbed by methane, so we are able to see them.

Another Moon

It was not until 1941 that a second moon was found orbiting Neptune. Gerard Kuiper was looking at photographs taken at the McDonald Observatory in Texas when he spied a dot of light on one of the pictures. Further study confirmed that it was a moon, which scientists eventually named Nereid, after a group of sea maidens in Greek mythology. Nereid has the most elongated orbit of any satellite in the solar system. Imagine that Nereid is traveling around a giant hula hoop in space. The hoop is not round, though. It looks like somebody sat on its side and squashed it into a long oval instead. It takes Nereid almost a full Earth year to travel around Neptune on its egg-shaped path.

Were there also rings circling Neptune? Since the other gas giants have rings, scientists figured that Neptune might have them too. Planetary rings are probably made up of the debris from moons or comets that were smashed in collisions. There is a great deal of debris in Saturn's wide rings. The beautiful, jewel-like bands of color, which were first identified as rings in 1656, are easily visible with binoculars or a telescope. The rings around Jupiter and Uranus are much thinner and fainter. These rings were not discovered until the 1970s. They have far less matter in them than Saturn's rings do.

In order to find out if Neptune had rings, astronomers watched through their telescopes as Neptune passed in front of a distant star. Six times between 1981 and 1985, they measured the light from the star as Neptune moved across it. Just before the planet's passage, light from the star blinked several times. This led scientists to think that Neptune had faint rings that briefly blocked the star's light from reaching Earth. The light did not blink on both sides of Neptune, though. Scientists concluded that the rings were probably not complete like the rings around Saturn. Maybe they were **ring arcs** instead. In order to find out more about Neptune, a more close-up look was needed.

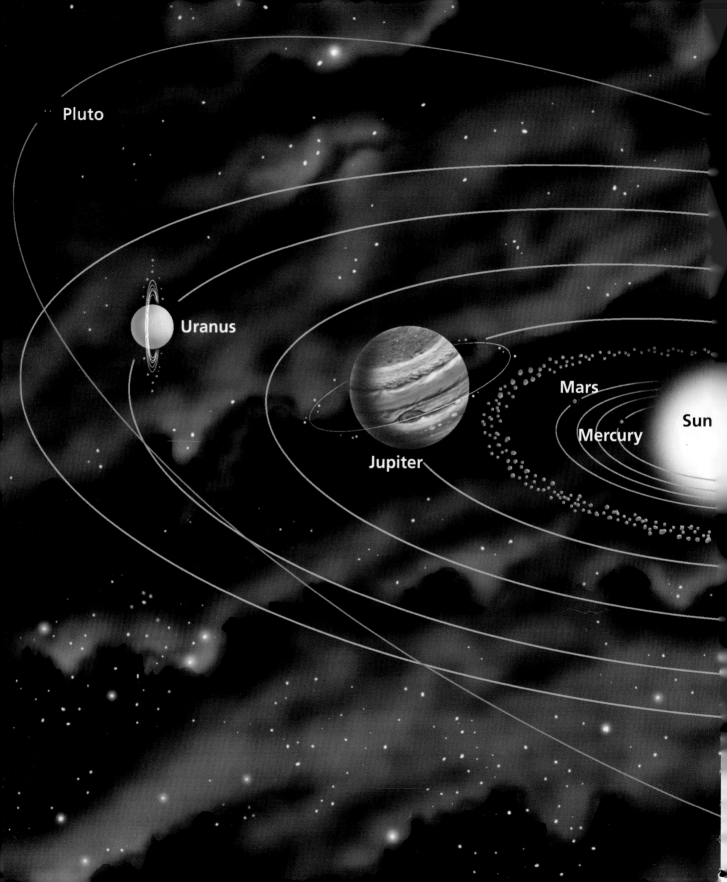